Cricut for Beggin...

A Step-by-Step Guide To Master
Your Cricut Machine and Design
Space and Make Money With It
How To Start a Business With Cricut
Can You Make It? Tips and Tricks
Included

Carol King

A little space to be creative

Table of contents

The information in the following pages is broadly considered a truthful and accurate account of facts and as such, any inattention, use, or misuse of the information in question by the reader will render any resulting actions solely under their purview. There are no scenarios in which the publisher or the original author of this work can be in any fashion deemed liable for any hardship or damages that may befall them after undertaking information described herein.

Additionally, the information in the following pages is intended only for informational purposes and should thus be thought of as universal. As befitting its nature, it is presented without assurance regarding its prolonged validity or interim quality. Trademarks that are mentioned are done without written consent and can in no way be considered an endorsement from the trademark holder.

Introduction

The invention of the Cricut device certainly transformed the booking & removing sport of craft. With its rotational cutter, this new tool can cut detailed symbols and patterns from various printing methods as well as provide a tailored, efficient, and quick, easy-to-make touch.

With paper, contemporary Cricut could still do incredible stuff, although they are capable of doing much more. The combination of the machine and the tools of the crunch model allows you greater artistic flexibility than it has ever been, and that you can choose a wide variety of materials for your Cricut with a huge assortment of tool attachments. This device is important for its simplicity and efficiency of use for both artisans and lovers.

There are many numerous types of crunching devices, they operate with various textures and enable various layouts.

Chapter 1: Getting Started with Cricut

At the center, Cricut's machines are super awesome printers. Functionally, they die cutters and artistic planners that help you bring fun designs together for different things you want to make. There are a lot of models out there, and there are many outstanding kinds to pick from.

There is a software called Cricut Design Space in the Explore series of computers, which helps you build anything you want to build in space, and then simply print it out. A Cricut machine might really support you if you're sick and tired of having the same pictures every day or if you're looking to cut out a pattern in vinyl without pulling your hair out.

1.1 Overview of Models

There are four major types of machines from Cricut, all of which are used to cut out different designs. The Legacy machines, including the Cricut Expression, the Cricut Expression 2, and the Cricut Explore, are still available. The Cricut Cuttlebug can also be found, but this is a machine mostly used for embossing and die-cutting, and this item has been discontinued as of spring 2019.

You've seen many incredible crafts that your buddies have made of Cricut on Facebook. You were mesmerized by the designs of your favorite Instagrammers. More than a hundred awesome Cricut designs you've pinned on Pinterest. But now you've chosen to take the offer to Purchase a Cricut! Yay! We're all so crazy for you! The only concern now is: which one to purchase? The Joy Cricut, the Explore Air Cricut 2, or the Cricut Maker?

The first thing to remember is that the Cricut Maker's all that can be achieved is The Cricut Explore Air can do that, and much more. If money is not a concern, then we firmly recommend that you purchase the Cricut Maker. However, let's be truthful, for many of us, money is a concern. So, what was the contrast here between Air 2 Explore Cricut Maker and the Cricut? Let's make a contrast between the two machines, should we?

The Cricut Maker

Cricut Explore Air 2

Cricut Joy

Cricut Explorer VS Cricut Maker

CRICUT CAPABILITIES	EXPLORE	MAKER
Simple to learn to use	✓	✓
Cuts, writes, and scores 100 materials	✓	✓
Cuts, writes, and scores over 300 materials		✓
Bluetooth capable	✓	✓
Two tool holders	✓	✓
Setting for 2 times faster cutting and writing	✓	✓
10 times more cutting pressure		✓
Compatible with Cricut's Adaptive Tool Set		✓
Rotary Blade for cutting fabric		✓
Knife Blade for cutting thicker materials including chipboard and basswood		✓
Scoring Wheels for creating creases perfect for paper crafts		✓
Engraving Tip for creating engraved handmade items		✓
Debossing Tip for creating debossed designs		✓
Wavy Blade for creating wavy edges quickly and seamlessly		✓
Perforation blade for creating evenly spaced perforated tear-offs		✓

1.2 Talk of the Town Cricut Joy

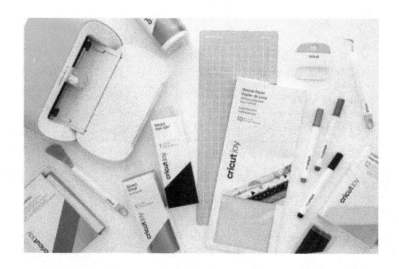

The youngest member of the Cricut cutting machines community is Cricut Joy. It is the tiniest intelligent cutting machine by Cricut (gauging just 8.5 inches wide, 5 inches deep, and 5 inches high), making it extremely compact, easy to transport, and beautifully portable. While this little fellow has some real state-of-the-art capability that will encourage you to build like no other machine in the Cricut family, never let her tiny size deceive you.

Cricut Joy also slices and writes, like most Cricut devices, but uses an entire host of instruments and components specially built for its limited footprint. All for Cricut Joy is well little but most significantly, purposely built to minimize fuss and irritation, thus optimizing convenience and seamless performance, ranging from vinyl's & papers to iron-on and card kits.

Anything for Cricut Joy, to be transparent here, is specially curated pens, blades, pads, fabrics, charging cable. While other items that you might have on hand can be cut down to match Cricut Joy's mats, with Cricut Joy, any Maker/Explore knives, mats, instruments, pens, etc., cannot be used. Often search for "Cricut Joy" specifically marked on the package to ensure what you're using with compatible.

Until now, it was sufficient to stick some kind of material (paper, vinyl, iron-on, etc.) to an adhesive mat and afterward load the mat into the machine to cut on a Cricut machine. If the slicing design were completed, you would unload the pad, extract the material from the mat, and clean it (very often).

Cricut Joy, indeed, replaces the whole operation. Underneath the feed manuals, it includes 9+ triggers that permit Cricut Joy Smart Materials to be cut without even a mat. Close to how you will install a molding machine, you literally feed these particular materials straight into the machine, and the template would be cut with the same accuracy as any other Cricut machine.

1.3 What Is the Adaptive Tool System for Cricut?

The Cricut Adaptive Tool System is unique to the Cricut Maker and, due to the Quick Swap Housing System; it helps you switch out razors without adjusting the housing. This not only makes it easy to build multi-dimensional designs but that also allows them more feasible. You can only obtain its other blades and points as long as you have one instrument with the housing. It certainly adds up to the average of $44.99-$49.99 for housing + tips, in which you can purchase only the pointers for $24.99-$29.99.

The Blade Wavy

For designs, the wavy blade brings a new meaning. To practically any content, you can apply a quirky edge and build a better look for gift bags or tags, envelopes, invitations, paper flowers, as well as a whole ton more.

Stuff to Remember Whenever the Wavy Blade Is Used:

This blade is like a rotating blade where tiny models and sharp corners actually happen. It's best to make sure that the cut is at least 3⁄4 percent.

The Maker does not monitor where even the waves are going to cut, and every cut is going to be different.

You will see a periodic cut line in Design Space, but you'll see the word WAVY below the line when you switch it to WAVY.

Components that Can Be Used

- Acetate

- Crepe Paper

- Fabric

- Vinyl and Vinyl Iron-On

- The Felt

- Paper

- Cardstock

Blade of perforation

The perforation blade allows puncture cuts that are perfectly separated. It makes it easier to create and use items like raffle tickets, countdown calendars, tear-off projects, and more. Without any need for folding beforehand, the perforation cuts make it so that you can break off instantly, and the cuts are smooth and tidy.

Debossing Tip

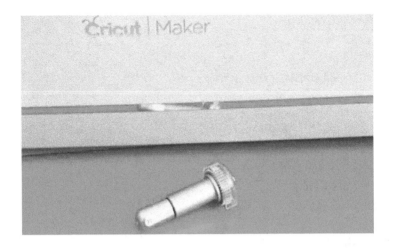

If you were sad to see that the Cricut Cuttlebug is no longer accessible, so this tip is perhaps the most exciting. The debossing tip could be used to apply professional-looking black anodized versions to invitations, envelopes, home decor projects, and more (debossing is where all the pattern is pressed back into the material, embossing is how it is lifted up). You will choose the location you want and emboss the label to create a completely unique and different design, unlike directories with an embossing/engraving device.

It can be used with the following materials:

- Cardstock

- Faux Leather/Leather

- Poster Board

- Kraft Board

- Paper

- Basswood

Engraving Tip

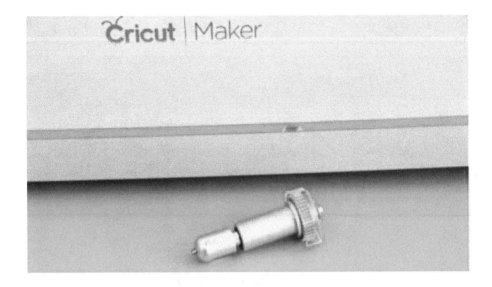

It has been something for a long period of time that we know many Cricut crafters have indeed been waiting for! A device that, without breaking or harming, engraves designs onto fragile surfaces. And, actually, the engraving tip is here! Monograms, patterns, adornments, and more can be engraved on the most fragile materials, including acrylic, aluminum, and more!

It can be used with the following materials:

- Cricut Metal Sheets

- Aluminum

- Acrylic

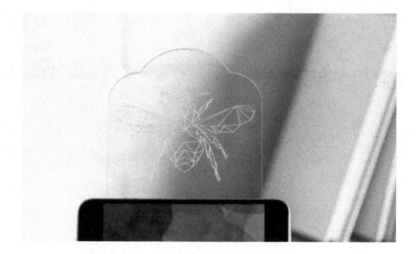

1.4 Unboxing Your Cricut Machine

So, you choose the Cricut, which is perfect for you. You went to the craft store to pick one up or eagerly waited for it to land at the store. The big day is here! Ready to have your brand spanking new Cricut unboxed! Are you thrilled? We feel that the owners of Cricut are in one of two categories:

- ✓ The individuals who plunge straight into their Cricut and unbox it as quickly as practically possible.

- ✓ The individuals who have been holding their Cricut for weeks in the box, who would like to start crafting more than anything, but feel perturbed.

- ✓

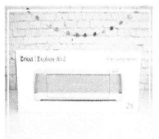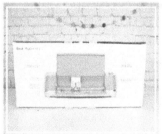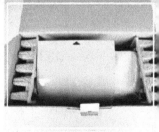

We've got your back no matter which group you fell under! Unboxing your Cricut and set it up for the first time! We're off to take you through every unboxing move so that you are all ready to get out and begin having a great time with your Cricut!

You will have the option of buying either the unit or a machine package when ordering your Cricut Maker or Cricut Explore Air 2 from the Cricut page. Most craft shops (not bundles) market only the unit. The Cricut will come in a box irrespective of whether you want a machine or a machine's bundle so that each box will come within the same basic elements. There are:

- Your Machine for Cricut Cutting

- Mat cutting

- Power Cord

- USB cable to link Cricut to your computer

- The Pen

- Cutting blade (It may be packed within your Cricut, so don't worry until you see it straight away).

- Get Started Packet (It will contain an instruction manual, a step-by-step manual to set up the new machine, tools for a new venture, and a Cricut Access trial membership.)

Let us just start unboxing your machine right away!

Stage 1: extract all from the box and ensure all the parts for your machine are there. For the list of parts that arrive with your machine, check the side of the box.

Stage 2: take your Circuit's plastic cover.

Stage 3: access the packet "Let's just get going."

Stage 4: switch the Cricut around, and the power cord is plugged in.

Stage 5: go over to cricut.com and pursue the Windows, Mac, IOS, or Android directions to connect your Cricut.com/setup.

Stage 6: you're fully prepared to do it now! Your Cricut is coming with a perfect first-timer project! As one of your first projects, we would really like to recommend that you make some bling for your Cricut! It's straightforward.

Chapter 2: Overview of Cricut Design Space

Are you prepared to start using Cricut? You can't even cut a task using your Cricut cutting machine without ever using Cricut Design Space, yet your design concept is fairly straightforward or sophisticated, and diverse.

If you are a techie, that can seem insurmountable to think about having mastered a new software program. For two purposes, you're in luck:

✓ Circuit's software program is incredibly easy to use when you understand the process, and when you're using it, you could even build on more advanced methods.

✓ Throughout your machine schedule, Cricut will automatically prompt you to install Design Space. Tap on the Cricut C application icon to launch the program when you have it downloaded, so let's get to it.

We would go through utilizing Cricut Design Space on your computer in this chapter of the book. As an app for iOS and Android users, Cricut Design Space is also accessible. The expertise between all is quite closely related to the desktop site and the app variant of Design Space, so users just might have to keep poking around to find out where every other tool resides.

2.1 Cricut Access

Cricut Access is a premium service sold by Cricut, which provides you with fast, unlimited access to over 100,000 images, over 100 fonts, and thousands of prepared designs. Members with Cricut Membership often enjoy a discount on Cricut products, like purchasing on Cricut.com materials and approved photographs from Design Space.

Memberships in Cricut Access are open, charged monthly or yearly. It includes a free monthly Cricut Connectivity membership when you buy a Cricut cutting machine that offers you a nice chance to check out the product.

You could still unlock photos, fonts, or primed designs since you're not a Cricut Access user, but to use them, you'll have had to pay a nominal, one-time charge.

2.2 Generating an Empty Canvas

Once you press on the New Project tab, you will be led to the blank canvas of Cricut. You can see three lines lined on board of each other on the upper left side of the panel. If you tap here, a list will come up.

2.3 Main Menu

✓ New Features

✓ The Country You Live In

✓ Help

✓ Sign Out

✓ Feedback

✓ Manage Custom Materials

✓ Update Firmware

✓ Account Details

✓ Link Cartridges

✓ Cricut Access

✓ Settings

✓ Legal

✓ Home

✓ Canvas

✓ New Machine Setup

✓ Calibration

If you do ever move far from your canvas and can't seem to figure out how to get back, just press this menu to switch to the landscape you were operating on by choosing Canvas.

Within this menu, the other groups you must involve:

Calibration: use this option when using the Print and Cut feature and also the Cricut Maker Knife Blade to adjust the device's blade for even more precise cuts.

Link Cartridge: it would never have to be used by the typical Cricut customer.

Menu option: so you'll want to recognize this if you're the fortunate owner of some aged Cricut cartridges! Press this tab, insert the specifics of your cartridge, and that you can now use the cartridge images inside the Cricut Design Space.

Settings: you can select whether the landscape has a complete grid, a minimal grid, or no grid under such a menu element. If the dimensions are displayed in metric or imperial, or where you want your job preserved for offline usage, you can still pick.

Design Space settings

Canvas grid

◉ Full Grid   ◯ Partial Grid   ◯ No Grid

Units

◉ Imperial   ◯ Metric

Saving for Offline

◉ Cloud & Computer   ◯ Cloud Only

2.4 Toolbars in Cricut Design Space

Toolbar Left-Side

The toolbar on the wing is where a significant sort of thing happens. It's divided into seven sticks.

Templates: you sometimes have a project idea that sounds amazing in your mind; however, once executed, you really don't know how it would feel. Reach Templates from Cricut!

New: anytime you would really like to establish a new canvas, press the tab. Cricut only allows you to work with one canvas at the moment; before even making a new one, ensure you save the existing canvas.

Cricut has produced illustrations with some of the most famous blank spaces. Select a blank, put the layout over the edge to get a taste of what the end product is going to look like.

Images: this section lists all great options in Cricut Designs Space, like images available to buy, images used in Cricut Access, and images uploaded.

Projects: this button pulls up a range of ready-to-make projects. To nail down the kind of ventures you're searching for, use the Class and Query boxes in the right upper toolbar.

Upload: in our experience, among the essential buttons in Design Space is the Upload icon. You'll have to learn how to use the feature if you'd like to cut your original artwork or cut files that you locate outside of Design Space. For about every task in this book, we refer to the upload button. It is so relevant which, in this book, we think that deserves its own portion. Let's give us more, then, should we?

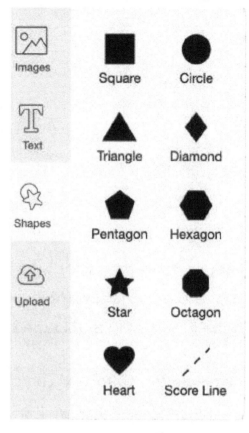

Shapes: Cricut offers ten simple shapes to its users that are 100 percent free to use. A cube, circle, triangle, diamond, pentagon, hexagon, star, octagon, heart, and score line are included. Under the Shapes List, you can find these simple shapes. You will also find alternate hearts, triangles, circles, etc., underneath the Shapes menu provided with Cricut Entry or available for a one-time price, whether you are searching for certain shapes with a little more elegance or pizazz.

Text: text is the place to start whenever you want to change comments to your canvas! With such a bit of help from your top toolbar, you will also want to fine-tune your text.

Top Toolbar

Since all the major things happen in the left-side toolbar, the upper and right toolbars are where the perfect fun starts!

Select/Deselect: using this command, select and deselect all the cut items on your canvas easily. Clicking and keeping your mouse down on one corner of your screen and dragging a box through all the things you want to pick is another way to select a cut product. In the right-hand toolbar, you can also press on each cut part you would like to pick when retaining the Shift button on the keyboard to pick it.

Undo and Redo Arrows: you should use buttons to undo and redo improvements to the pictures on your canvas that you have made.

Fill: this box is used when you use your home printer to print components. "No Fill" means that you're only going to cut that piece. "Print" means you're going to print and then cut.

Print Type Box: once you switch the fill type to Print, you can pick any colors or styles for the Print and Cut feature from such a box to fill in your cut file.

Linetype: use this option to pick whatever you want each of your shapes/images to do with your Cricut. Where, when you're using Cricut Explore. Cut, Draw, or Score would be your tools. Split, Draw, Score, Engrave, Deboss, Wave, and Perf. Are your choices at this time for the Cricut Builder? This list may expand as more Scalable Resources are published. Your choices will be Cut and Draw for the Cricut Joy.

Material Colors: pick different colors for each cut feature from this box to help maintain which image will be cut straight from each color background.

Size: using this frame, resize your cut unit by adjusting the boxes in width and height. The standard is to remain proportional to your picture, but if you want to adjust this, you can press the Lock button over the "H" to activate ratios. By clicking the icon and using the arrow box, which exists in the bottom right-hand corner of the image, you can also scale an image. For that, Press Lock, which resides in the bottom left-hand corner of the picture when it is picked, to enable the dimensions when sized this way.

Rotate: to rotate your file, use this button. When attaching score lines to a project, we use this key regularly.

Position: while using the Position box, shift your image to precisely wherever you like it on your board. Trying to position your image at 1X, 1Y will place your picture on your landscape 1 inch down and 1 inch over. Even so, this amount does not correspond with where the picture on your cutting mat will be sliced. After you press, Make It while you edit the mat, placing your image on the cutting mat can be finished.

Align: you'll have to pick two or three cut components from your canvas to use the Align Feature. With this method, you can arrange left, middle horizontally, right align, top align, center vertically, bottom align, center, horizontally disperse, and vertically disburse items. We always use this tool directly until attaching, welding, or chopping parts.

Arrange: Space Design functions in layers. The first feature that you apply to your canvas will become the bottom layer, and the top layer will be everything you add after it. Often you want an aspect to stay at a higher level than where this is situated. Using the Organize button to use the Send to Back, Move Backward, Move Forward, and Send to Front choices to move components all the way to the other side, back one layer, forward one layer, or all the way to the front. This is another technique that we sometimes use directly prior to connecting, welding, or slicing sections.

Flip: this tool enables you to switch horizontally and vertically your pictures.

Edit: you can find Cut, Copy, and Paste choices underneath the Edit box.

My Projects: tap this icon to link all the projects you have stored in Design Space to a registry. To go back from this monitor to your canvas, tap Canvas on the hand side of the page.

Machine Button: from this dropdown list, select the type of machine you're running (Cricut Maker, Cricut Explore, or Cricut Joy) to unlock functionality unique to a certain machine.

Make it: if you recognize only one Design Space button, that's the button! When you are able to start cutting your task, tap Make It.

Save: ensure that by pressing the Save button, you will reopen your project some other day. If you're using items from inside the Cricut Design Space, you will be permitted to save your file publicly so that you can exchange the file with friends. You would only be authorized to save files for individual use when using any fonts or photos you have submitted.

Text Toolbar

A text toolbar will display under your top toolbar until you've applied text to your canvas. (Note: only when your text is chosen can you see it.) The options used with this toolbar are:

Style: you can render the font bold, italic, bold italic, writing, then back to normal again under the Style menu.

Curve: it was one of the most important features of the Circuit. To give your words an arc economically and efficiently, interact with the slider under this box.

Letter Space: to place your letters closer together—and farther apart, use this feature.

Line Spacing: render the gaps between terms clustered closer together—and further apart on top of one another.

Advanced: you can ungroup the terms under this menu, making the spacing between all letters and lines simpler to play with. For individuals using script fonts to aid link letters, this tool is particularly common. Ungroup to Letters, ungroup to Row, and Ungroup to Layers are the three alternatives available inside this menu.

Alignment: change the font so that it is oriented, aligned left, or aligned right.

Font: as you push this icon, it opens up every font accessible for you to use, including your computer's Cricut fonts and fonts. By pressing the All, Framework, and Cricut keys, you can sort your own fonts against fonts owned by Cricut. There is also a font search box by name and a filter box that helps you filter fonts by types, like My Fonts, Multi-Layer, Single Layer, Blogging, and Offline Saved.

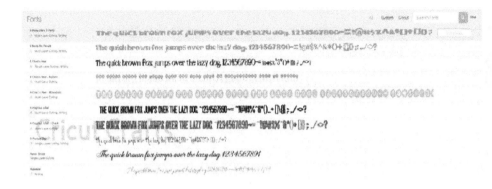

Font Size: to adjust the numerical value of the font size, use the font size box.

Toolbar Right-Side

We've got the right-side toolbar, last and not least. Realizing how to use these techniques correctly would make it so much simpler to work in the Design Room. In this section, the tools include:

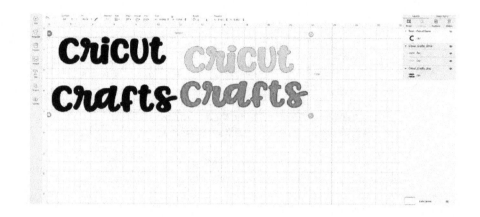

Ungroup: to ungroup a collection of photos that have been clustered together, use this icon.

Duplicate: by clicking this button, duplicate any design feature you have chosen.

Delete: by clicking this button, delete the design feature you have picked.

Group: in your venture, pick two or more photos and click the group button to band them all. Why would you want this to be done? Moving clustered pictures together on the canvas is smoother. If you get the dimensions for your style worked out, if your images are clustered together, it will instantly resize the other images it's grouped with when you resize one image, holding it in the same proportions.

Slice: inside Design Space, Slice and Weld are two of our favorite features. Say Slice is like a cutter for cakes. Lay one design element on top of another, pick all parts, choose Slice, and the top design will be cut out of the bottom design by the Design Space. We laid the words Cricut Crafts on top of a heart in this series of images for starters. We then picked the two pieces, pressing Slice, and I'm left with the words cut out of our heart by Cricut Crafts.

Weld: if Slice is like a cutter for cookies, Weld is like glue. We are using the Weld button to attach quickly two or more design features into one cohesive object. For, e.g., we want to link the heart to the end of the "s" in Cricut crafts in this series of images. We move the heart to the place we want it to be, pick both the craft and the heart of the Cricut, and then press Weld. The two pieces are then made into one solid piece by Construction Room.

Attach: to group objects of the same color, this button is used to cut the way you have them spread out on your cutting mat. For instance, let's assume that we want these four hearts to be cut such that 1 inch is equally spaced between each heart. If we space them on our canvas and don't attach them to make the most effective use of your cutting material, they will cut right next to each other.

However, if we pick and attach all four hearts, they'll cut the same way they're spread out on our canvas from the cutting pad, making them easier to move to your project.

Detach: you can see a Detach button appear where even the Attach used to be before you have attached cut components together until you press on them. To remove your attached parts, push this button.

Flatten: this button is used to convert components to an image in Print and Cut. For e.g., we have three distinct design features in this set of photos: a white circle, aqua phrases that say Cricut crafts, and four pink hearts. If we're going to press, Render It as it is, three separate mats will cut the pattern (one white, one aqua, and one pink).

Nevertheless, if we pick all three design features and press Flatten, it will transform the design into one picture, which will be printed and then cut with your Cricut using your printer. The Cricut will cut out of the loop around the edge and leave the remainder untouched.

2.5 Make It Magic!

In Design Space, we have covered all the design resources, but what happens after you press Make It? Let's go over the simple steps to get the idea to be cut.

You will be guided to a page that highlights one of your cutting mats on the left-hand side of the page until you press Make It. The first slicing mat on the monitor will also be seen to be huge. You will also have two choices under each mat on the left-hand toolbar:

1. From a dropdown list, you can pick your material size.

2. To duplicate your image, you can press the mirror slider. Mimicking is an important aspect when dealing with materials such as iron-on and insoluble ink, which are sliced with the right side face down on the mat. If you don't mirror these products, the images/words will be in reversal when you add them to the blanks.

You also have the luxury of shifting the template around wherever you want it to be on the mat when you have a cutting mat picked. When you use scrap bits of cutting material, or when you use your inlay tip to etch on metal, this function comes in handy.

Three dots (...):

When you press on the "..." button, two choices pop up. Shift and conceal artifacts when you press on the Move Object. It will give you the option of shifting your design to another cutting pad when you just want to make the most out of your cutting material and pack because most cuts as practicable into one mat; this will come in handy. When you press Hide Object, the layout is covered so that it doesn't cut anymore.

Circle Arrow

You will be able to rotate the template with the circular arrow. While you are struggling to use up the last single meter of your cutting stuff, this feature makes life easier.

You can press until you have your designs put on the cutting mat. Then you will be taken to a computer where you will pick your cutter and material for cutting.

On the projector, your preferred cutting materials will light up. Click Browse Products if you'd like to choose a cutting product that is not displayed. This will put up a comprehensive list of materials for cutting categorized by type. To make finding the material you are searching for easy. Also, there is a search box.

If you find that you want your template set on a different location on the mat once you get to this screen, or if you have forgotten to mirror the design, you can press the Edit button underneath the cutting mat that you want to edit on the left-hand toolbar. A wider image of your mat, which you can modify, would come up. There is also a reflection slider at the base of this box, so you can adjust the style to be reflected.

All Materials   Favorites | All Categories ▾ | Search All Materials | 🔍 |

Art Board

Cork, Adhesive-Backed

Corrugated Cardboard   ⟳

Flat Cardboard   ☆ ⟳

Foil Kraft Board – Holographic   ⟳

Foil Poster Board   ⟳

Kraft Board   ⟳

Light Chipboard – 0.37 mm

Metallic Poster Board   ⟳

Cardstock

Cardstock (for intricate cuts)   ⟳

Glitter Cardstock   ⟳

Heavy Cardstock – 100 lb (270 gsm)   ★

Medium Cardstock – 80 lb (216 gsm)   ★ ⟳

Fabric

Burlap, Bonded

Canvas

Denim, Bonded

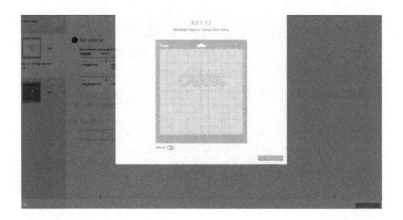

You will be instructed to load some special knives, instruments, or pens into your Cricut once you have chosen your cutting material and to put your cutting mat into your machine.

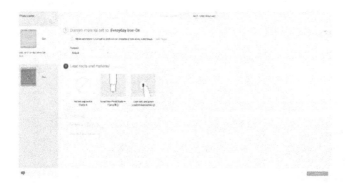

Simply click on the field that says Custom Material Applied to, and it will open up the box where you can pick a new cutting material if you ever need to adjust the cutting material you chose.

SnapMat

Cricut has a feature called SnapMat that is available exclusively via its app on phones and tablets. It helps you take a picture of your cutting mat with the material spread out while you use the SnapMat tool, and then put your cut anywhere you want it to be on the mat. When cutting fabric with a large, bold pattern, we love to use this feature. It helps us to "fussy cut," which suggests removing your form from a particular part of your cloth pattern.

And you've got it there! Space 101 in Cricut Architecture! We hope this acted as a fantastic preview for you and that in the background, you heard about a tool or button or two that you didn't even know about. The further you get Design Space, the more instinctive it will feel, and you'll be designing projects like a professional before you know it!

Chapter 3: Projects Using Most Famous Materials

Now is when the fun really gets off! Uh, projects! With your Cricut trimming unit, you can make a million and one stuff. Our purpose is to include your insights and talents that you can then pass to develop your own creative ideas in this book. In the following fields, we have divided the programs.

- ✓ Fabric

- ✓ Chipboard and Basswood

- ✓ Faux Leather and Leather

- ✓ Infusible Ink

- ✓ Vinyl

- ✓ Paper

✓ Vinyl/Iron-On Heat Conversion

✓ Special Materials

For each material, you can find 4 to 6 projects. We have attempted to use a huge spectrum of blanks for iron-on, acrylic, and Infusible Ink (mugs, wood posters, clipboards, tops, coasters, etc.) such that you get to really understand how to use each product as well as how much you could do for them!

Projects with Paper

Before you even learn how to use your Cricut, the paper is a wonderful medium, to begin with, because it was one of the least costly materials you can find. Many card-makers purchase Cricut's to help push their card-making talents to the next level, not intending it for the remainder of their crafting lives to open up horizons!

Some suggestions for designs on paper:

- ✓ Planner accessories

- ✓ Luminaries

- ✓ Shadowbox art

- ✓ Bookmarks

- ✓ Cards

- ✓ Gift tags

- ✓ Paper flowers

- ✓ Paper succulents

- ✓ Scrapbooking embellishments

- ✓ The Cricut can cut paper in all different thicknesses, finishes, and weights.

3.1 Thank you Gift Labels with Love!

Machine: Cricut Maker or Cricut Explore Project

When you own a Cricut, you can find that you're included in buying so many items you can now produce, including gift labels! To create lovely gift tags, utilize your Cricut to cut pretty, painted cardboard, photo book paper, obsolete calendars, old cards, and another sheet! Or use simple white cardstock and the Print and Cut function of Cricut to create a massive pair between your Cricut and your printer, as we are going to do in this project!

Items:

- Light Grip Cutting pad

- Blade fine-tip

- Cardstock White

- Color Printer

Directions:

Phase 1: log in to Cricut Design Space and import the cut file gift tags following Cut File Upload's directions in this book.

Phase 2: by tapping on the design then using either the arrows that reside in the lower right-hand corner of the design or the Scale tool in the top toolbar, size the design so that it is 9.25 inches on its widest side.

Phase 3: please press Make It. Follow the on-screen instructions. Next, you'll use your printer to print your gift tags. You will then pick your slicing material, put your printed gift labels on the cutting pad, and use the arrow button on the machine's right to mount it into the machine. Watch for the Cricut C to begin blinking until your mat is prepared, click the key, and your machine will start to cut. Click the arrow again as your machine stops cutting, and it will unload your mat.

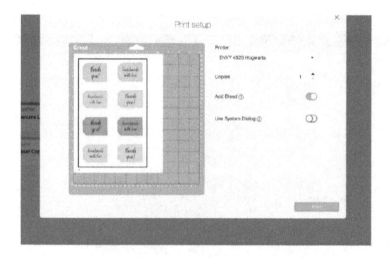

Note: don't ever be distressed if the Cricut doesn't automatically start cutting. You're definitely going to find a light switch on above the blade. When the light scans back and forth over the paper, the mat will pass. This is the black boundary that your Cricut is looking for because it knows where to cut.

Stage 4: extract the gift tags cautiously from the cutting pad. They are ready to use!

Enjoy your lovely tags for presents! What other ways would you think of decorating presents with your Cricut?

3.2 Illuminated Sleepy Town

Machine: Cricut Maker or Cricut Explore Project (Cricut Joy Compatible)

Residential luminaries are easy-to-make. These paper homes, illuminated with a rechargeable batteries flame, produce a lovely impact for any mantle or as a centerpiece of a table. Using one cut file to render houses to build the tiny village of your dreams in a range of sizes and colors!

Items:

- Stylus Scoring or Wheel Scoring

- Dots with Adhesive or Glue

- Battery-operated candles

- Village Sleepy Cut File

- Cutting pad LightGrip

- Blade fine-tip

- In a range of sizes and colors, cardstock (We used 8.5 x 11

- For our luminaries, 12 x 12, including 12 x 24)

Directions:

Phase 1: sign in to the Cricut Design Space and import the Sleepy Village cut document.

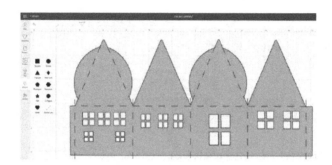

Phase 2: on the right-hand toolbar, press the Shapes button and pick the score line. For all the positions shown in the photo, attach score lines. There are all the ways you'll like to be able to slide your paper to make your homes. Pick the home and then all the score lines, and press Attach until you have all your score lines put wherever you want them.

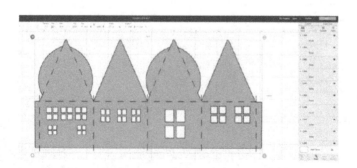

Phase 3: to build as many houses as needed, press the Duplication button. To unlock the size ratio, tap the lock button. Carry with the house's sizes, trying to make some taller and skinnier, a few really wider and shorter, some larger and some smaller. Pay attention to the amount of cutting mats and paper you have so that each house suits your paper.

Phase 4: please press Make It. Follow the on-screen instructions. On your cutting mat, rest cardstock and place your scoring stylus or scoring wheel into your machine. Using the arrow button on the right of the device, push your mat into the machine. Wait for the Cricut C to begin blinking once the mat is prepared, click the button, and your machine will continue to score and slice. Click the arrow again as your machine stops cutting, and it will offload your mat.

Phase 5: after all of the houses have been split, fold the cardstock at each score line. To keep the tabs in place, use Sticky Dots or adhesive.

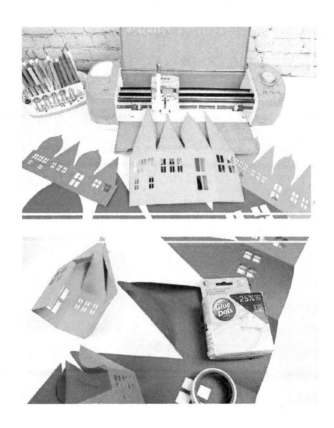

Within each home, put one or more battery-operated candles and build your own lovely, sparkling, sleepy town!

3.3 Making Kawaii Thumbnails Using Magnets

After growing sick of having textbooks lying open on desks, the sofa, and almost every other portion of our home, we developed the inspiration of these magnetic thumbnails! Fold the bookmark over the tab where you left off, and the magnetic tighter will firmly mark your position on the back of the section! Such magnetic bookmarks make tiny crafted presents the loveliest! We gave them to our local craft fair as stocking stuffers, presents for friends, and even sold them!

Machine: Cricut Maker or Cricut Explore Project

Items:

- Scoring blade or stylus scoring

- Paper with magnetic sticker

- Scissors

- 8.5 x 11-inch white cardstock

- Color Printer

- Cutting pad LightGrip

- Blade fine-tip

- Cut File of Kawaii Food Magnetic Bookmarks

Directions:

Phase 1: sign in to the Cricut Design Space and import the Kawaii food cut files.

Phase 2: by tapping on the design, using either the arrow in the bottom right-hand corner of the layout or the Sizing tool in the top toolbar, scale each bookmark as required. Our bookmarks had a length of 6 inches.

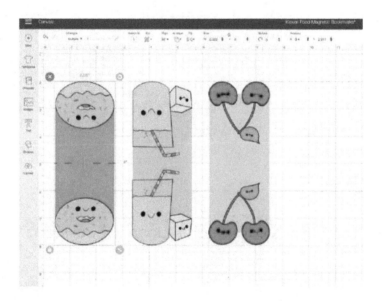

Phase 3: attach your bookmark with score lines. On the left-hand tooltip, tap on the Shapes icon. Click the line with the ranking. Rotate 90° along the score line. Place the score line at the top of each bookmark, then pick both the bookmark and the score line; under the Align tab, press the Center Vertical choice. Click the Connect button for both the score line and bookmark picked. For each bookmark, repeat.

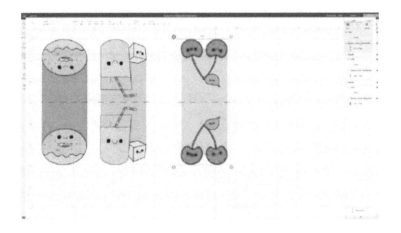

Phase 4: please press Make It. Follow the on-screen instructions. Next, you'll use your printer to make your bookmarks on the cardstock. You will then pick your slicing content, put your printed page on the slicing mat, and use the arrow button on the machine's right-hand side to load it into the machine. Wait for the Cricut C to begin blinking when your mat is prepared, click the key, and your machine will start to cut. Click the arrow again as your machine stops cutting, and it will offload your mat.

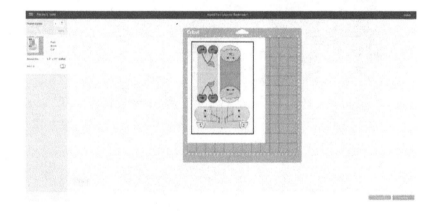

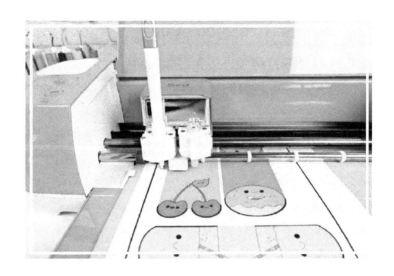

Note: don't be distressed if the Cricut doesn't automatically start cutting. You're definitely going to find a light switch on above the blade. When the light scans back and forth over the paper, the mat will pass. This is the Cricut who's really waiting for the black border to see where to cut it.

Phase 5: take the bookmarks cautiously from the slicing mat and fold each one from the score sheet.

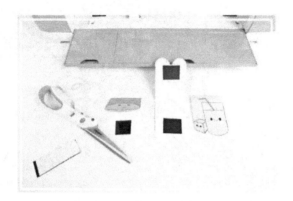

From your magnetic stickers board, cut two squares of magnets and place the magnet from the inside (non-printed side) of the bookmarks so that both magnets are lined up. Drop over the page of a book with these cute Kawaii food bookmarks and tag your spot!

3.4 Parent's Day Cards

This design is to show you how and when to create an envelope and a picture card. To convert this into a Mother's Day or Father's Day card, we've provided modules. When you understand the process, you could use this cut file by inserting your own additional adornments to render picture cards for every celebration or special event. Christmas, Valentine's Day, commemorations, graduations, weddings, wedding shower thank-you cards, and more are perfect for picture cards!

Machine: Cricut Maker or Cricut Explore Project

Items:

- In your favorite colors or prints, three pieces of 12 x 12-inch cardstock

- Photo Card File Cut

- Blade fine-tip

- Scoring blade or stylus scoring

- Dots with Glue

- File split envelope

- 8.5 x 11-inch white cardstock

- An embellishment of the Mother's Day Print and Cut File

- An embellishment of the Father's Day Print and Cut File

- Color Printer

- Cutting pad LightGrip

Directions:

Phase 1: sign in with the Cricut Design Space and import the image card, envelope, then cut files for Mother's Day or Father's. Be sure to save components of Mother's Day and Father's Day components as an illustration in Print and Cut.

Phase 2: scale the card such that the width of it is 5 inches. With curved edges, scale the square picture backer such that it is 4 inches wide. Measure the envelope to be 11.75 inches long. Scale the components of Mother's Day and Father's Day into the perfect sizes to decorate your card.

Phase 3: to remove the letter and the packet liner, use the Detach button on the right-hand toolbar. The envelope, packet liner, and card incorporate score marks. Tap on the Shapes button in the wing toolbar to attach the score line. The Score Line is picked. On the card or envelope, put the score where you'd like, then pick both the score line and what you add it to and press the Add button from the toolbar on the right-hand side. To focus your score line on the card, you will want to use the Align button at the top toolbar.

Phase 4: please press Make It. Follow the on-screen instructions. Next, you can print components of your Mother's Day or Father's Day using your printer.

You will then pick your slicing material, put your printed pages on the slicing mat and use the arrow button on the right of the machine to mount it into the device. Wait for the Cricut C to begin blinking once your mat is mounted, click the key, and your device will start to cut. Click the arrow again as your computer stops cutting, and it will unload your mat.

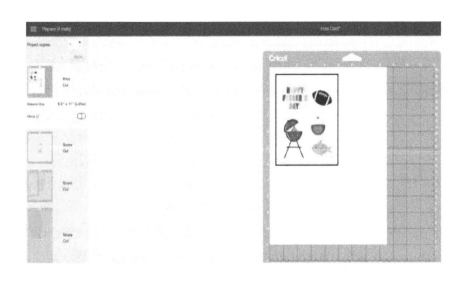

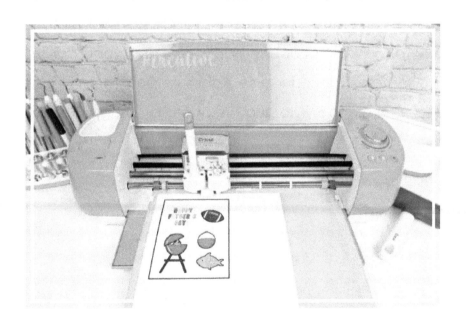

Note: don't ever be worried if the Cricut doesn't automatically start cutting. You're definitely going to find a light switch on above the blade. When the light scans back and forth over the paper, the mat will pass. This is the black boundary that your Cricut is looking for because it knows where to cut.

Phase 5: fold each component with a score line all along the score line until all the pieces have been removed. Then, tape the envelope lining to the envelope's interior using Adhesive Dots, then glue the wings of the envelope closed. Next, attach items from Father's Day or Mother's Day to the front of your card. Eventually, place a frame into the card gap and use Adhesive Dots to secure the picture in place by adding the photo backing.

Chapter 4: DIY Fabric Projects in Cricut

Just about every fabric you toss at it, The Cricut Maker will cut. With the Cricut Maker, a few of the preferred fabrics to cut are:

- ✓ Cotton Quilting

- ✓ The Felt

- ✓ The Canvas

- ✓ Fleece

- ✓ Mink

- ✓ Flannel

- ✓ The Denim

- ✓ Fabric Sequined

- ✓ oilcloth

4.1 Sloth Sleep Mask

This cute little sleep mask is an adorable personalized gift and is ideal for sleeping parties to wear. Create a whole sleepy sloth sleep mask family and give it to everyone you love!

Machine: Cricut Maker Project

Items:

- Blade fine-tip

- Vinyl heat conversion (we used brown, black, and white for this project.)

- Protective Layer Cricut Iron-On

- 10 x 12-inch piece of cloth for your mask's front (we used quilting cotton)

- Easy Press Cricut or Iron

- Computer stitching

- 10 x 12-inch cloth piece for the rear of the shield (we used fleece)

- 12-inch piece of elastic black

- Thread in color matching

- SCISSORS

- Clips or pins from Wonder

- Sewing Needle

- Cut file of Sloth Sleep Mask

- Adhesive cutting pad with Fabric Grip

- Regular cutting mat grip

- Rotary Blade Cricut

Directions:

Step 1: sign in to the Cricut Design Space and import the Sloth Sleep Mask's cut files.

Phase 2: by tapping on the layout or using either the arrow in the lower right-hand corner of the design or the Size function in the top toolbar, scale the design so that it becomes 8.5 inches tall.

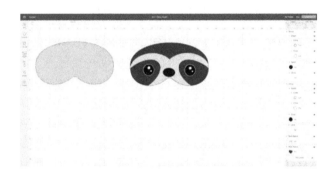

Step 3: please press Make It. Follow the instructions for cutting each sleeping mask piece, loading the needed materials, slicing the blade, and using the Fabric Grip sticky slicing mat and the rotating saw; the sleep mask-shaped sections can be cut out of fabric.

By using the standard grip cutting pad and fine-tip cutter, the sloth face parts can be cut out of the HTV (please ensure you load your HTV with the glossy colored side down).

Using an arrow on the right-hand side of your Cricut, fill your slicing mat with each content, pick the necessary material from the drop-down menu, and then mount the cutting mat into the unit. Look for the Cricut C to begin blinking once your mat is loaded, click the trigger, and your computer will start cutting. Click the arrow again as your computer stops cutting, and it will offload your mat. For each material that you are cutting, repeat.

Step 4: it is time to add the iron-on forms to the front of your sleep mask fabric once you've cut all the bits. It requires stacking, which can be challenging when the iron-on does not want to incinerate. We notice that blending is so much simpler with the Cricut Iron-On Protective Mat. For the sort of HTV you are using, review the Easy Press Heat Setting Guide to see what temperature and how long each piece gets pushed.

Initially, add the sloth's hairline and eye patches, accompanied by the whites of the eyes, the black eye sections, the white-eye dots, and the nose.

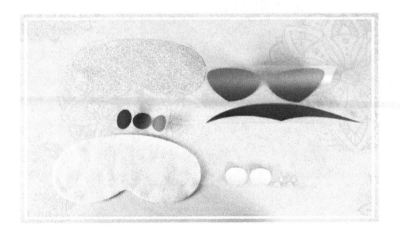

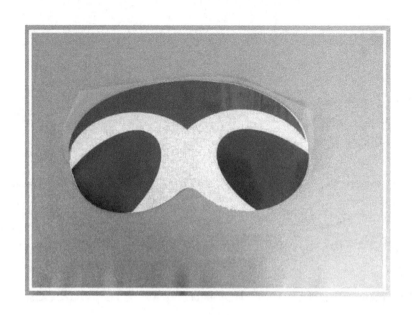

Phase 5: it is time to sew now! Pin your elastic on the sleep mask's back flannel part. On either foot, let the elastic ceiling the edge of the mask by 1/2 inch.

First, you would like to create a "sleep mask sandwich." Facing up ought to be the side of the fleece with the elastic attached to it. Lay the front of your facemask on top of it with your eyes and nose facing down. As iron-on has been added to the mask's face, we consider using wonder clips so that you do not poke a hole in your iron-on and/or be strategic about where you put your pins.

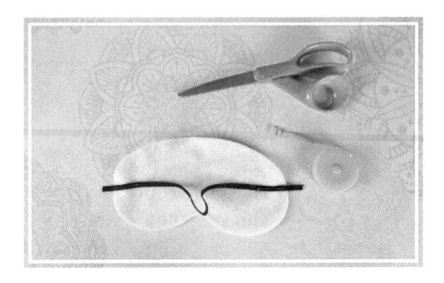

Step 6: using a 1/4-inch seam allowance, stitch all the way on the outside of your sleep mask and make careful to remove a 1-inch or wider space to turn your mask right-side out.

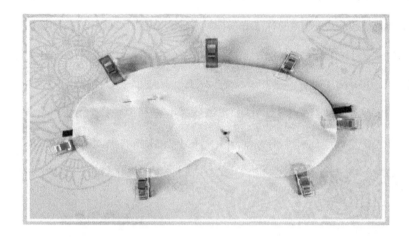

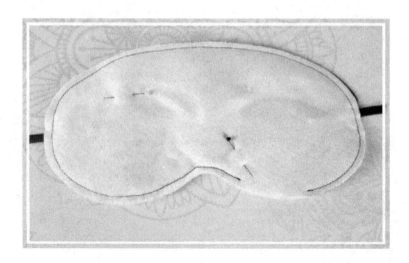

Step 7: turn the right side of the mask out, press in place (make sure you use the iron-on shielding sheet while squeezing the face of the sloth), and stitch the hole you used to transform the right side of the sloth mask using a needle and thread.

And you've got it there! A cute Sleep Mask from Sloth! For a promised sleepy sloth sleep, wear it!

4.2 Animal Collar Bandana

For animal lovers, or dogs themselves, the Animal Collar Bandana allows perfect presents. Create them and create your furry buddy a supermodel in a lot of fun designs and colors!

Items:

- Sewing machine

- Scissors

- Cricut Rotary Blade

- Wonder clips or pins

- Thread in coordinating colors

- Fabric Grip adhesive cutting mat

- Over-the-Collar Pet Bandana cut file

- Measuring tape

- Fabric

- Cricut Easy Press or iron

Directions:

Phase 1: sign in to the Cricut Design Space and import the Pet Bandana cut file.

Phase 2: weigh your pet's collar where the bandana would like you to rest on the collar. Press the cut file, and the region at the bottom of the triangle (the region here between two arrows in the photo) will be resized to a certain dimension. To make the second piece it is the same size, press the Duplication button.

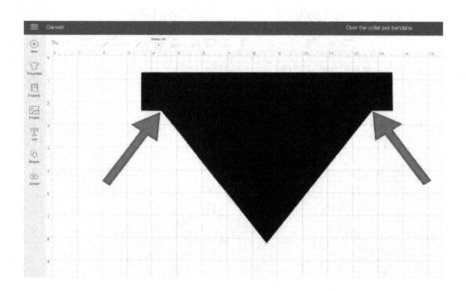

Phase 3: please press Make It. To pick the kind of fabric you're using, obey the guidelines. On the Fabric Grip pad, put the fabric. Use the arrow button on the Cricut Maker to mount your slicing mat into the device, and afterward click the blinking Cricut C. Continue for the second material piece.

Phase 4: from the cutting pad, clear the cut cloth. Next, we would like to fold over the ends of the labels so that as we slip the bandana over the collar, we do not have bare cloth edges. Fold the tab on top so the tip meets up with the triangle's base. Once more, fold, so you have a clean, tidy trim, and use your Easy Press or iron to press into place. Repeat on all bits of cloth with tabs 1 and 2.

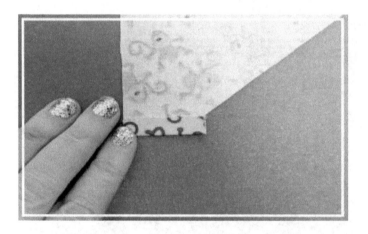

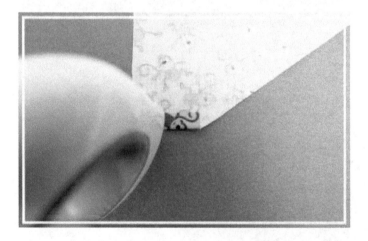

Phase 5: using a 1⁄8-inch seam allowance, sew down each hem.

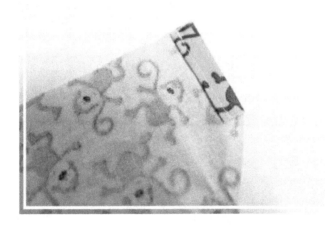

Phase 6: after that, with the printed side of the garment squished in the center, line up the corners from both pieces of your bandana. By using a 1/4-inch seam allowance, thread over the bandana's top lip. Next, stitch down to the bottom of the triangle from the base of your hemmed tab and back up to the base of the second hemmed tab. Make sure to leave open the hemmed tabs; this is how you will flip your right-side bandana out and feed the collar through it.

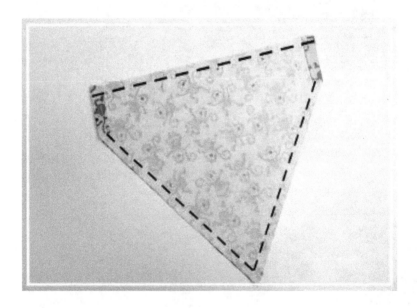

Phase 7: flip your right-side bandana out, making sure all your corners are pushed out. Iron your scarf so that it's clean and bright on the corners.

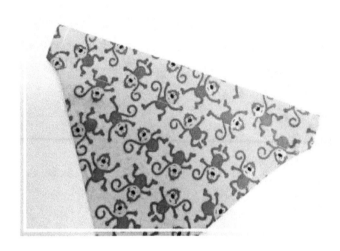

You've got an adorable over-the-collar pet bandana now! Feed the collar of your animal through the bandana and put it on your pet. This is such a cute way to let the personality of your pet really shine!

Conclusion

So, that's the highway's limit. If all the ideas and techniques given in this guide have been learned, then cheers! In order to make sure that you get better at the earliest, you can hold on with this, before you can deal with things.

However, you must press high after learning the stuff. Keep this masterpiece close to you for purposes of reference.